HYPERSPHERE ...

A JOURNEY AT THE SPEED OF GEOMETRY

Revised Edition

Robert B Cronkhite

Copyright @ 2012
Robert B Cronkhite

All rights reserved. No part of this book may be reproduced in any form, except for the inclusion of brief quotations in a review, without permission in writing from the author or publisher.

Library of Congress Control Number: 2012914356

Edited by Susan Reyes Morrison
and David E Morrison

Printed in the United States

Dedicated
To my wife, Sarah
And to my granddaughters,
Alyssa & Kylie

In Memoriam

I would like to acknowledge the contribution of my father, Robert Cronkhite to this work. His encouragement throughout childhood and since was astounding. In my far younger days with my paternal grandfather (Robert B. Cronkhite) trying to figure out the possibilities of horse racing results, he challenged me to conceive and build a machine one day with only one moving part. It was our mutual interest in amateur and shortwave radio that astoundingly would lead to the idea of an oscillation of matter (with matter itself now the signal) that finally led to the concept, experimentation and the machine herein called the Hypersphere.

Though he had only a partial formal education up to the eighth grade, came from a laborer's family of the rug mills of Amsterdam, New York and having been a welder at General Electric Company in Schenectady, New York, he did not hesitate to wonder and intrigue me.

It is ironic that despite the common and limited perceptions of our present state of 'civilization', that his supposed 'dementia' was really impractical to some and yet an unlimited new window for us upon entanglement and the edge of physics. His new perceptions were a marvelous aid to my own 'inconvenient thinking and wonderings'. All this allowing us to just use what has been so systematically ordained by those who decide such things, not to use: that *one* as a prime number, the practical use of the *imaginary numbers* and to just consider Hyperspace as a reality enough to engage geometry as mechanism.

Thank you Dad for your courage, despite your discouragements, to dream and try to achieve what has been so quietly embedded within the earlier dissertations of others, and significantly with me.

As Sir Isaac Newton had remarked, we concur that we too 'have seen so far' for we also 'have stood on the shoulders of giants'.

God bless you Dad and thank you for your valuable efforts in our work.

Your son, Robert Bruce Cronkhite

Third Level Machines

Hyper Geometric Machines

According to the *World Point Formula*, $ds^2 = dx^2 + dy^2 + dz^2 - dt^2$, a third level machine is time like, dependent upon the coefficient of ds^2. When the coefficient is positive the machine is space like; for example, automobiles, airplanes, and people walking the street. When the coefficient is 0, then the machine is light like; for illustration, a radio, a laser, a light bulb and other photon-based apparatus. When the coefficient is negative, then the machine is time like; presently, the passive clock, but tunneling and entanglement will be included here. All machines always have some element of the other two. What makes the machine able to be categorized is where it functions predominately.

Third level machine experimentation has been going on for quite some time (since the later 20th century when my father challenged my thinking of a one moving part machine, my own intrigues with relativistic and quantum machine, and motor possibilities, and then in the early 21st century with the contemplations of Noninertial mechanisms).

It became apparent that with a third level machine there would be more profound repercussions. In studying the nuclear explosion, where some of the mass is converted into energy (which in another paper I relate to the equivalence of information, as information flow is one of the features of the active Inertial Geometry we are embedded in) as the nuclear explosion reaches a third level of function. This is when the mass is converted into energy (swiftly traded with hyperspace for energy).

A third level machine is capable of inertial to Noninertial back to inertial function. Thus, what we observe as tunneling, entanglement and super luminal effects are that, the effects of phasing for an inertial particle or photon to a Noninertial state on this side of the speed of light (*c*), to the other side and back; traversing suddenly large amounts of space and time, and then just as suddenly, not.

Starting in 5758/1998 AD theory slowly became experiment and from 5765/2005 AD experiment allowed for observation and measurement. Quantification as well as qualification of not only nano and microscopic effects, but also macroscopic effects of the third level mechanics. Hyper Geometric Mechanics would be proper.

These are examples starting with the most simple to the more complex -- *The Temporal Diffraction Grating* with the "Where is NOW?" experiments; the *Natural Geometric Computer* analogy of a fistful of sand is a more particle and macroscopic illustration (in another paper); the Hyperplane, where an artificial event horizon affecting information flow was utilized; as a culmination and much more complex, the Hypersphere, where articulation for experiments in entanglement and tunneling for macroscopic size objects entailed.

Third Level Hyper Geometric Machines
A Natural Computer

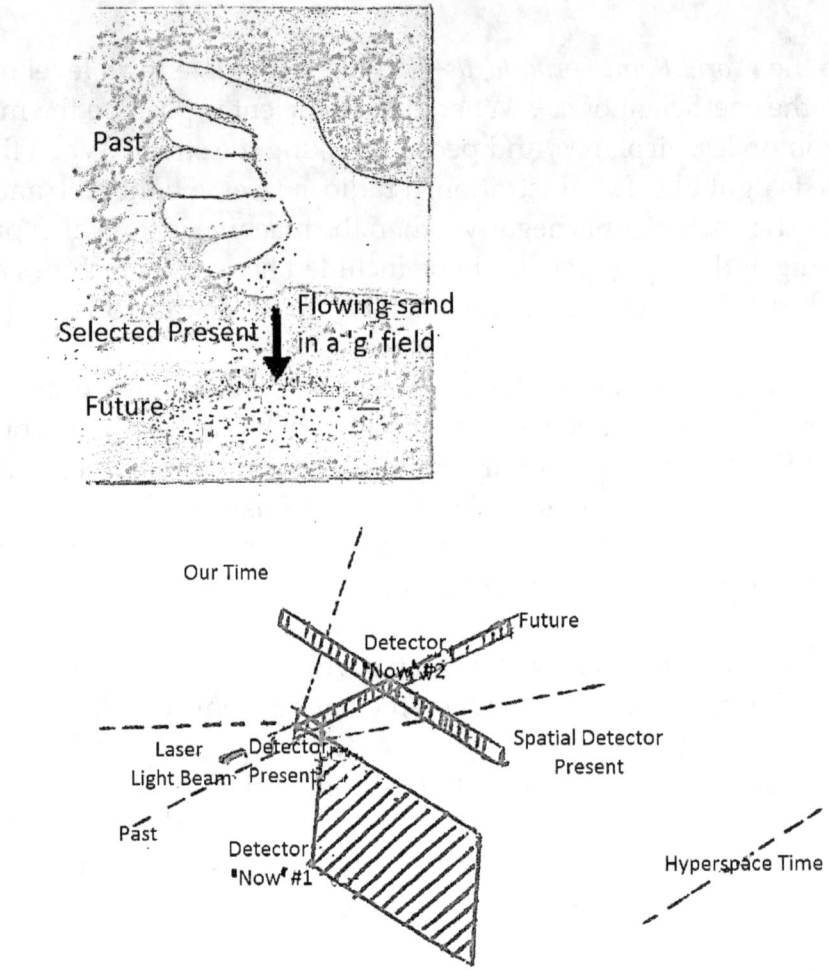

(it is at a right angle to a commonly used *spatial diffraction grating*)

These two machines are not only illustrating spatial concepts (where most concerns have been focused up to the present, with minor in temporal considerations), but are also demonstrating very major temporal observable and measurable effects; especially the *Temporal Diffraction Grating* which is able to quantify even the past, present and future in our commonly experienced forward time, but also reverse time. As Maxwell's *equations* allude to in his concept that every electromagnetic has a forward in time (retarded propagation), but a reverse time (advanced propagation) that appears much weaker. Essentially the present as we experience the sequences of events, linear and Nonlinear, is the strongest signal above noise; then the future signal, at about $1/x^2$ to now; and the past before the present, at a very weak $1/\Omega_\aleph$ (*Omega Aleph* $\Omega_\aleph$).

The quantification of the past, present and the future, along the fourth dimension -- the Hyperplane using a radius that equals c (*the speed of light*) of the rotational velocity of a laser beam carrying a binary number sequence, like an artificial event horizon of a black hole; before the beam reaches the c radius, the rotational velocity is below c and the beam is sweeping clockwise as at the origin. When the beam is detected at c where the rotational velocity equals the speed of light, then the beam's presence freezes. Move the detector farther to a greater radius, then the beam is observed moving counterclockwise while the information still streams in forward time. At this point, while the beam appears to change direction, its information is still forward from the center out. But if the information is also entered according to the radial distance instead of through the center, one can see that some of the information is reversed in clumps, which depends on the amount of information and input along the inner radius that was used. Why? Because inputs 1, 2 and 3 each displayed their own '1', '2' or '3' on the shorter and below c radius, than at the c radius. Only one of the numbers would be frozen on, and at the longer radius it would read '3', '2' and '1'.

Even an image of the number as backlit from a common small lamp would do the same thing -- an interesting effect. Yet at all times if each input progresses in forward time, then there is a similar forward progress of numbers. If information comes in through the center or through the inputs along the inner ring radius at below c progressively, then the forward progression is always observed at the outer radius above the c ring detector. But small time reversed loop does seem to occur in limited four dimensional sections according to the limited four dimensional geometry able to be articulated in the Hyperplane.

The Hypersphere, the third hypergeometric machine is the articulation of inertia within the local Inertia Geometry, commonly known as 'gravity', which we term local curvature of Inertial Geometry. To articulate a macroscopic object to faster than light, it has to have no inertia. It has to be positioned outside of the local curvature and even the globe curvature of Inertial Geometry. As any particle has mass and momentum, and the massless photon also has momentum, so there are still inertial geometric effects. It is easily observed for the microscopic quantum world and within the Hypersphere, however not for the macroscopic world.

Here tunneling and entanglement are allowed for a large object that escapes inertial contrast, like the shadow or light beam rotating along with information, dependent on the circumference of the articulated radius. Initially starting with the small spheres of pyrolytic graphite then with small spheres of bismuth, both with diamagnetic properties, bismuth having a much better characteristic.

4

The Hypersphere   (Perhaps this simple third level machine with one moving part,
the oscillating mass, is one of the 'shapes of the things to come'.)

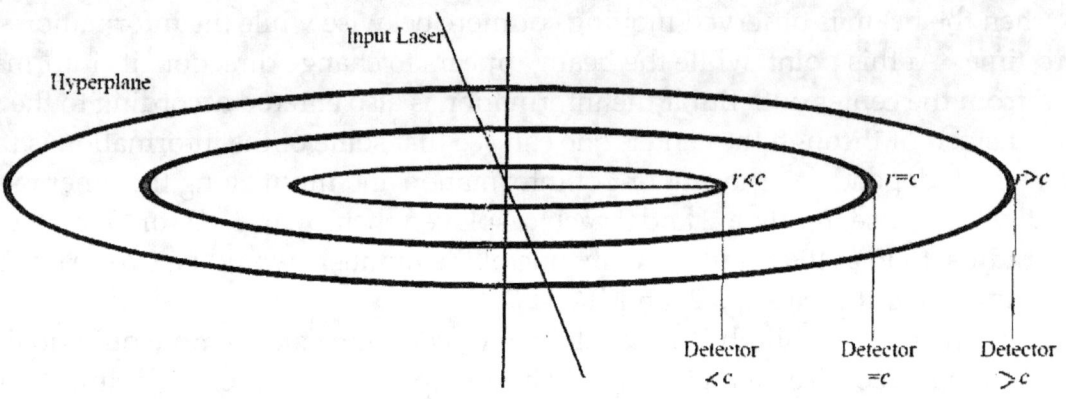

Hypersphere in Oscillation

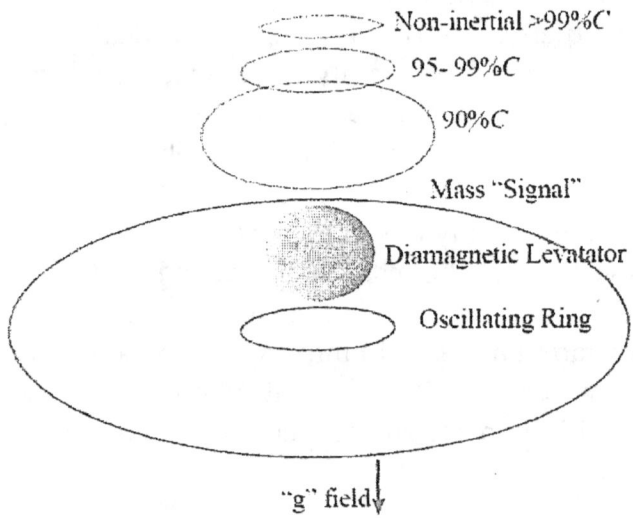

Before we get into the deeper analysis of the Hypersphere's operations, it is interesting to note its parameters and those implications that its operation elucidate.

The final velocity of the Hypersphere, which matches the speed of Noninertial Geometry, is very close to $3.7 \times 10^5 c$, much faster than the speed of light. In the Hypersphere the mass itself is the signal, as the signal riding an electromagnetic wave in radio. But in this case the mass escapes from the Inertial Geometry, which is an active curvature of space time and where the mass has inertial characteristics. Slowly the mass is oscillated at higher and higher frequencies and then gently in order to avoid an explosive interchange of energy for mass, an 'atomic explosion'; the mass slips from the local and even global Inertial Geometric of our space time and into Hyperspace time, which has reverse entropy, the same as reversed time.

The mass may now traverse greater distances in time and space, for it is coherent in superposition in both time and space, and able to tunnel and entangle to and from its oscillating driver ring, as long as the driver ring maintains the proper environment for these endeavors (just as quantum microscopic particles would, due to having very low to no inertia while in a proper Noninertial environment).

Initially the only difference is the volume and mass of the macroscopic signal. The pendulum and Newton's *Cradle* (where he demonstrated the law of conservation of momentum) anticipate to oscillation of mass in a gravitational field. The period of the oscillations are dependent on the strength of the gravitational field; as so is commonly done, a pendulum is used to measure local variations of the Earth's gravitational field geographically.

Essentially we can add to our third level machines utilizing inertia. The pendulum, the gravity driven clock, Newton's *Cradle* and passive clocks predominately used for the measurement of the passage of time are all again third level machines. Certainly it is written, "There is nothing new under the sun".

Basically the Hypersphere is a quantum/relativistic harmonic oscillator. The ring is the secondary of a transformer and magnetic amplifiers. It is also an active clock, unlike the passive ones commonly used to just measure time. The Hypersphere affects time as well as space.

Some or all of the cosmicscopic superluminal observations may well be categorized as Noninertial as well. They too follow the *formula of relativistic aberration.*

$$tan\frac{\theta'}{2} = tan\frac{\theta}{2}\sqrt{\frac{c-v}{c+v}}$$

Because of the angle in some cases, as like a shadow with rotational velocity of a beam of light, when at superluminal speed it offers Noninertial observations.

Even Hyperspace, which is the inversion of our experienced 'normal' space time, is like time space. Light speed seems to be the border, the meniscus between our space time and hyper time space. Though in Hyperspace, they seem normal and we are the odd ones. Our space time is forward and entropic, so too is Hyperspace; but our time and theirs are flowing in opposite directions. All of this is in our rotating and expanding universe black hole that is four dimensional.

An illustration of our space-time, the speed of light border or meniscus and Hypersphere

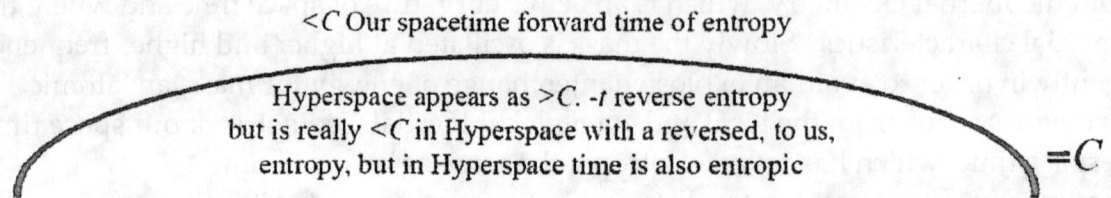

To us we are looking at the outside of the sphere and for Hyperspace they too see the outside surface of the same sphere. Yet each of us uses the other space time for a shortcut, consistent with entanglement and tunneling. To each observer, the other externally is Noninertial, because on the other side of the speed of light border seems 'faster' than light. But what seems to be happening is that high speed of this time reversed relative to the other and appearing 'Nonlocality' are just two different Inertial Geometries each with a reversed to the other direction of time flow and direction of the entropic sequence of events flow.

In the small quantum world this reverse causality does not mean much, and on the macroscopic and cosmic scales the distances in both superposition of time and space as being more than one time, more than one space at the same time and the same place are not in any causal conflict, thus affirming Hopkins *Causality Protection Conjecture*.

For example at near $3.7 \times 10^5 c$, 100 light years, 100 years ago and 100 years into the future, and at the same time as hovering over the ring are too well separated to affect any local causality problems. Essentially, it is not WHERE Alpha Centauri *is*; it is more WHEN it *was, is* and *will be* ..... or just WHEN *it is*. So if I send a light pulse to Alpha Centauri and it takes 4.3 years to get there and if I hypergeometrically transverse the same in 3 minutes Earth time, and send a signal back and arrive back 3 minutes after that, my sent signal will arrive 4.3 years later. Both signals will arrive as the first, then the second, 8.6 years apart. But I traversed within the same in approximately only one half hour.

For though in the hypergeometric Noninertial hyperspatial traverse I had traveled both time and space much faster, was better tunneled and entangled in superposition spatially and temporarily, my quick traverse was also my being at Alpha Centauri and above the ring at the same time, in the same place; and in the other times at the same place and both places at three different times; in my half hour still above the ring and at Alpha Centauri, within the half hour on earth and 4.3 years ago and 4.3 years in the future! Remember, this is the superposition as experienced in quantum mechanics already, but now macro- and cosmoscopically. So what about the causality? Each light signal will take 4.3 years to go one way and I would not yet have received the first one and only just transmitted the second one. I would receive the second signal 4.3 years later. So locally, both for my space time on earth and for anything in the local space time on a planet around Alpha Centaur, there will not be any paradox of causality. Causality is protected in the superpositioning of BOTH space and time.

According to the Young *Double Split Experiments* and those like them, an object, because of its own wave nature can occupy more than one place at the same time. (I must add the corollary that an object can also occupy more than one time at the same place and also occupy more than one time and more than one place at the same time and place), hinting at Deutsche's *Many Worlds Theory*, but now made clearer than what at first appeared. It is entanglement and tunneling further extrapolated to macro and cosmic scales that have up to the present been accepted without question for the nano-, microscopic and smaller scales. This may be why an inertial electron and an inertial proton in the confines of atomic space time do not act inertial. The electron whether spinning around or somehow oscillating as a standing wave, which is similar to the Hypersphere's basic physics, becomes Noninertial. There is no more entropy. The atom does not cease being an atom over time, begging the question, "Where does the energy come from"? The Noninertial suffers no entropy, can exceed the speed of light, experiences entanglement, tunneling and super-positioning of both time and space.

So the speed of Noninertial geometry is near $3.7 \times 10^5 c$. Light is on the border and without mass and only momentum experiences no time, superpositioning in time and space, so relative to itself and capable of entanglement and tunneling; the electron and Carbon (C^{60}) can too because of so little mass and thusly so little inertia. At that scale the wave nature is greater than the particle nature; under the proper physically articulated environment so can macro scale objects and on the far greater scale, cosmic objects when near relativistic velocities particularly and nearer extreme gravitational environments, like the Event Horizons of black holes. Again considering 'gravity' as really the active Inertial Geometry, the speed of light on the 'border' of Hyperspace, the other side with only a reversed entropic time flow of events, then entanglement, tunneling and superpositioning are attainable and understandable. The Noninertial is just past light on our side and just past light on the Hyperspace side. It is in this Noninertial part of the Inertial Geometry where "the next frontier beckons".

The Hypersphere   (Abstract)

The Hypersphere is a third level machine with predominant temporal effects. It is basically a harmonic oscillator. This is the signal of mass of a magnetic amplifier. It is the secondary of a transformer that is capable of very high frequencies and adjustable wavelengths. It is also a Nonlinear accelerator and at its extreme of use approaches to near 90% to 99% of the speed of light after entering inertial geometric hysteresis and becoming Noninertial geometric. At that point it can suddenly be at the speed of Noninertial geometry, which is 3.7×10^5 of the speed of light. Based on its characteristics, "the rocket is obsolete".

In the 12 years from theory, to experiment and refined experiment, it was discovered that first with a diamagnetic particle of pyrolytic graphite and later with the better diamagnetic particle of bismuth, that a mass could be oscillated and articulated in opposition to the local curvature of inertial geometry, more commonly defined as a gravitational field. They are maintained in opposition to each other and form a relativistic harmonic oscillator, very much like a quantum harmonic oscillator.

The mass as the signal suffers increasing Lorentzian effects until within 90% to 99% of the value of c (the speed of light). Suddenly in this turbulent area, Lorentzian effects subside and as suddenly, the speed of Noninertial geometry is achieved at $3.7 \times 10^5 c$. Relativistic tunneling and entanglement effects suddenly are prevalent as inertia is escaped, for the mass has slipped from the local curvature of inertial geometry, or more commonly the local gravity.

This macroscopic particle is now in a relativistic state of entanglement and tunneling, as quantum particles with very little mass and thus inertia very often experience. The mass signal is now capable of articulation in multiple spaces and multiple times, superpositioning and coherent with itself across progressive distances in the two opposing directions of space motion and temporal motion, all at the same place and the same time.

A small sphere of bismuth or pyrolytic graphite, hollow and containing micro and later nano scale sensors, was attempted. The micro circuits were achieved to present and were able to sense temperature and cosmic rays and transmitted by entanglement, advanced and retarded low wattage (QRP) in milliwatts telemetry.

It appears that with oscillations lasting approximately 15 minutes the mass was coherent, entangled and tunneled, one light year between the areas just above the driving ring and a distance of one light year into interstellar space, one year forward and one year prior to the present of the experiments in time ….. all at the same time in the same place as where and when the experiments took place.

If the period of the experiment had extended to near 2.4 hours, then the space and time differences would be respectably 100 light years away, 100 years backward and forward from the *where* and *when* of that particular experiment.

(More is to follow as experiments are continuing and extra local effects near the place of experiments are noted --- particularly the slowing of clocks next to the ring and of lesser slowing rates at increasing distances from Amsterdam, New York to Latham, Waterveliet and Verona all in New York.

The intensities followed the *inverse cube law* like tides do. It was tidal and temporal Lorentzian effects decreasing away from the oscillations.

Also of note is how at the transition from inertial to Noninertial, an image echo of the mass and an after image slowly rose and faded above the driver ring. It followed the *relativistic aberration formula*, as shown below.

$$\tan\frac{\theta'}{2} = \tan\frac{\theta}{2}\sqrt{\frac{c-v}{c+v}}$$

HYPERSPHERE

A Scientific Dissertation at the Speed of Noninertial Geometry

Prospectus

1. Hypersphere: a third level machine
2. Inertial Geometry: gravity as only the effect of an underlying active geometry
3. Both forward and reverse time are mechanically relevant
4. Ωℵ (Omega Aleph) as a multistate constant and the primary prime number
5. Hypergeometric Theory of Machines

The Prospecti

As the concepts upon the four subjects are expounded, there will be the proposal considered that the universe is as Einstein and Gödel proposed, a rotating sphere. The further implication by this author is that it is a rotating black hole. The hole or sphere of our universe is four dimensional, so that everywhere its surface is accessible at the speed of light. This will propose again that though we do not witness any exceeding of the speed of light, we do see effects resulting in reversal of entropic flow (the direction of time), the quick tunneling of macroscopic objects at great distances and experimentally a speed of nearly 3.7×10^5 the speed of light.

Preface and Warning

The main user of intent is the State of Israel, since 5758/1998 AD to the present time of 5770/2010 AD of this manuscript's production are for their exposition of its profound implications. For some, the contents herein will possibly be disturbing. Its repeatable proclamation of a deterministic universe where in certain extreme conditions, even macroscopic particles and objects approach similar effects that have been observed and measured (with admittedly limited human measurement resolution) for micro and even smaller scopic scales.

With particular phenomena of reverse causality, tunneling and entanglement of now macroscopic scale particles and objects, when providing a proper extreme environment of inertial geometric conditions, these results further imply that Albert Einstein and Kurt Gödel were quite correct in proposing a spherical rotating universe similar to a spinning black hole whose surface is an event horizon.

With the event horizon equal to the speed of light, it is the four dimensional surface with each side having its own direction of entropy or time, which is the Nonlinear and sometimes linear sequences of events, locally and globally. On the other side of the surface is a reverse only in reference to the side of origin which appears as Hyperspace and has Noninertial elements, as well as inertial.

Hypergeometric Theory of Machines

The Hypergeometric Theory of Machines proposes that according to the famous formula, $ds^2=dx^2+dy^2+dz^2-dt^2$, any possible machine will only be one of three hypergeometric natures, based on the coefficient of the formula's answer for ds^2. Thus a third level machine is predominantly time-like and is mostly functional in Hyperspace

This is where the Hypersphere Machine functions predominantly. Following the curvature and even escaping it, it suddenly achieves nearly 3.7×10^5 of the speed of light.

All inertial elements --- those showing increasing Lorentzian effects as velocities approach the speed of light, and Noninertial elements that have exceeded the speed of light, which occurs suddenly to just under 3.7×10^5 of the speed of light --- is where tunneling and entanglement occur, as well as the profound, yet measurable and observable occurrences of now even macroscopic objects occupying more than one place at one time and more than one time at the same place.

(I will leave it up to the reader to discern and provide repeatable and verifiable experimental evidence and measurement, to quantify and qualify in order to prove or disprove what is herein this scientific dissertation.)

It is redefining quantum mechanics as Probable Mechanics and including macroscopic everyday objects with similar characteristics in extreme inertial geometric induced environments to allow their own slippage to a Noninertial state and sudden attainment of great distances equally in time and space.

All the essential parts of such third level technology have existed in similar and varied forms since the latter 19th century and the mathematics had origins in the latter 17th century. Poincare's *Conjecture,* Euler's *Identity Formula* and multi-dimensional geometries were implying such. With the early 20th century's observation and limited resolution measurement of quantum scale particles, it was assumed that all of this was normal, for so much had been counterintuitive to classical physics. The same is true with relativistic scales of velocities for particles. Both extremes appeared limited to such conditions.
"Welcome to Hyperspace time."

In essence, following 12 years of experiments (the latter three years intense) the rocket is obsolete on scales within the greater part of the solar system, especially on interstellar and pragmatically even to the near Milky Way Galaxy environment to include the Magelenic clouds.

Starting with particles of pyrolytic graphite and later bismuth (both diamagnetic materials), small spheres have already left Earth and appear to have macroscopically tunneled, and entangled on a macroscopic scale as well, and then have returned to be still hovering over the driver ring of the oscillating secondary of a specially designed *Hypergeometric Harmonic Oscillator*.

There are three more experiments that will refer to the previous two third level machines that were used to develop concepts and measurements leading to the Hypersphere. These are in successive order, a "Where is Now?" experiment that uses only a laser, a photoconductive glass and a series of small detectors, arranged as a *Temporal Diffraction Grating* at a 90° angle to a regularly used spatial diffraction grating in order to validate Maxwell's *EM Equations* that in the 19th century, equally considered the existence of not only the well-known retarded wave but also advanced waves in reverse time.

The second machine was the Hyperplane. It allowed the producing of what would be considered a 'shadow' event horizon. This is where rotation of a laser beam beyond a certain radius produces reverse movement in its procession, while its origin continues in forward procession.

Thirdly, another experiment used Sir Isaac Newton's *Conservation of Momentum Demonstrator*. Basically in a four dimensional universe, where each sphere touches is between two opposing clouds of the surface electrons, a quantumly small space. (It is here that I was able to get to the closest of foliation as to begin measurements for the Hypersphere's place to "slip through" to Hyperspace. Also in explaining to anyone where the edge of the universe is, all I needed to do was slip a piece of paper in between two of the spheres. In conjunction with the Temporal Diffraction Grating I would be able to demonstrate "Where NOW is!") This is the beginning of quantification of not only Hyperspace, but the fourth dimension.

1st Prospectus – Hypersphere

The Hypersphere is essentially a hypergeometric machine. Its ds^2 coefficient is predominantly negative. The sphere or mass signal is driven by a concentric magnetic ring that is oscillating with a magnetic field driving the diamagnetic sphere and keeping it levitated, but oscillating as a harmonic oscillator at increasing high frequencies.

At some point *relativistic aberration* effects in proportion to Lorentzian effects become more and more pronounced. The small sphere flattens and creates an increasing hum in the air. It appears reddened and even a slight luminance appears. Starting slowly, then quite quickly it apparently rises and fades away. At this point it has achieved the frequency of foliation and suddenly loses its inertia and is hurtling along at nearly 3.7×10^5 of the speed of light. As long as the driven element is oscillating, the sphere is entangled above the driver ring, yet not visible and is at the same time Noninertially very far away at the $3.7 \times 10^5 c$. According to calculations, in less than 15 minutes it has gone 10 light years, and not only 10 light years in space but back 10 years in time! Now expanding this to just less than 2 ½ hours, it will have traveled 100 light years in space and 100 years back and forward in time. But being farther away in both space and time, it has no local causal paradox problems. This would confirm Hawking's (Cambridge, England) *Casual Protection Conjecture*. If macroscopic objects can under such conditions traverse space and time with equal distance, there can be no local interference.

But a new question arises. If one were to traverse back to where the earth was 100 years ago and traverse both space and time again equally, then would local causality be affected? It is then not *where* Alpha Centauri is, but *when*. The greater the spatial distance (proportionally the greater temporal distance) and the *when* become more than equal in importance to *where* something is.

2nd Prospectus -- Inertial Geometry

In introduction and summary, the use of instruments that had not yet been invented or readily available, specifically the *Temporal Diffraction Grating* ("Where is NOW?") apparatus, the Hyperplane (for application of an artificial event horizon for Noninertial phenomena, particularly information flow) and then the Hypershere, and even the Newtonian *Conservation of Momentum Demonstrator* had to be made. These were needed for experiment and observation in an attempt to better understand what the possibilities were of macroscopic objects tunneling in time and space, being entangled in both time and space (as it seems to happen all the time to quantum scale phenomena); and to see what happens after something becomes Noninertial (seems to lose its mass and inertia) and exists in more than one place at a time and more than one time at a place.

This is where Inertial Geometry ends (or 'gravity' and inertia, as we know it) and the leaving of the local space time curvature for Hyperspace time; also where the macroscopic acts as the microscopic. These are extreme behavioral areas in our 4D universe. The speed of Noninertial geometry is 3.7×10^5; one light year in approximately one minute!

Basically all that we perceive as 'gravity' is the active inertial geometry of the rotating 4D universe we are embedded in. As long as we remain embedded, then we experience the effects of inertia, 'gravity' and forward time. Forward time is the direction of entropy we experience. Its rate as well as all sequences of events' rates is inertially geometric dependent. That is why the Lorentzian factor is so observable and measurable.

The surface or 'meniscus' of our 4D rotating universe is the speed of light. (It will seem a bit counterintuitive to some as we get into this, but it will make sense later.) For if I take the five spheres of the Newton *Conservation of Energy* or Newton's *Cradle* "toy", the surface of the 4D universe is comprehensible by placing a thin sheet of paper in between each sphere. For though spatially they appear at different places, they are in 4D space time at the SAME space time!

When we observe the sequence of rotation of a beam of information through a Hyperplane machine, we also measure and observe a 'shadow' of an Event Horizon, not too dissimilar to the Event Horizon of a much more influential massive black hole. Again it is only Inertial Geometry that is the common denominator. The beam of information shorter than the radius of c goes forward, when at the radius of c it freezes, and longer than the radius of c, it reverses! Thus with Newton's *Cradle* and with the Hyperplane we are observing and measuring the effects of Inertial and Noninertial Geometries and our rotating 4D universe. But that is not all.

3rd Prospectus – Both Forward and Reverse Time are Mechanically Relevant

With the Hypersphere, we continue to quantize more and more of our measurements and observations of 4D space time, and more and more of not only our space time, but Hyperspace out of its subtler presence.

Going back to the aforementioned *Temporal Diffraction Grating* we can quantify as well as observe forward and reversed entropic time flow, making some confirmation upon Maxwell's *Electromagnetic Equations*. For he had speculated that there would be evident not only retarded electromagnetic radiation as we normally observe, but also advanced electromagnetic radiation as well. The retarded waves are propagating in forward time, our well observed direction of entropy. But also there would be possibly a more weakly measurable and observable propagation of the same electromagnetic waves in reverse time, negative entropy, in a reverse direction of time and entropy.

It at first seems counterintuitive, but it is only the SAME WAVE in both FORWARD and REVERSE time. And with the *Temporal Diffraction Grating*, this becomes quantifiable and apparent. With this machine of measurement and observation we are again able to observe a subtle or 'shadow' event horizon effect. It seems that the two entropies, the forward and reverse of time, and the two observations of the SAME signal are only separated by the speed of light.

That surface or meniscus is the border between our space time and Hyperspace time. It is here somewhere within the border that suddenly the Noninertial speed of just under 3.7×10^5 of the speed of light occurs. There is a frequency and associated wavelength near .001 meters, where the foliation of the inertial 4D 'surface' is porous to the Noninertial; where not only common quantum, micro, nano and smaller fluctuations occur, but where macroscopic fluctuations can be articulated with the most shortened Lorentzian wavelengths for normally massive particles we are familiar with every day. This is near the Planck Length of 10^{50} area

In the Inertial Geometry to Noninertial, transition to Hyperspace inertial 4D with reverse time Euler's *Identity Theorem* seems relevant.

4th Prospectus – Omega Aleph $\Omega\aleph$

It has been the norm to not consider *one* as a prime number. It seems as absurd as *zero* and *infinity* to consider for the quantification and measurement of reality and like imaginary numbers left out of practical use.

But in our reconsidering these new perspectives, let us suppose *one* as a relevant prime number. Granted it is divisible by *one* and is equal to itself, so it does qualify as a prime number. But with the increasing relevance here in these pages of inertial and Noninertial geometries and Hyperspace, *one* as a prime number is very significant. Not only is it to be shown that 1 is a prime number, but also how its other faces like -1, square root of *1*, square root of -1, and *zero* and *infinity* are also related that they all equal each other while showing a different face under differing conditions of the curvature of space time, transitioning through event horizons and when any object, including macroscopic objects, are in more than one place at a time, more than one time at a place and undergoing entanglement at great distances of both space and time!

It will be argued that not only as *one*, but its other derivatives as exhibited, is important in the expansion of concepts of the Hypersphere, Inertial Geometry, and the Nonlinear Sequences of Events.

(This is further explored in another paper on Omega Aleph, Ωא.)

5th Prospectus – Hyper Geometric Theory of Machines

The last machine, the Hypersphere is in the spotlight as it is the observation, measurement and the resulting quantification of macroscopic objects being articulated to tunnel and entangle -- not only in more than one place at the same time, but in more than one time at the same place. Based on the idea of a quantum harmonic oscillator, it is actually a relativistic macroscopic harmonic oscillator. The mass being oscillated is a diamagnetic material such as pyrolytic graphite, later bismuth and then smaller amounts of such diamagnetic materials encompassing non-diamagnetic material, but of course with less efficient results from the same power inputs to the magnetic drive ring. The diamagnetic mass itself is the signal, similar to the signal of a radio transmitter. The circular magnetic ring of the driver, like in an antenna, is really the secondary ring of an oscillating magnetic amplifier. The magnetic ring drives the mass up against the 'gravitational' field, the local curvature of the local inertial geometry.

The local 'gravitational' field (the local curvature of the inertial geometry) pulls the mass down. The mass or 'signal' is articulated to oscillate faster and faster with increasing frequency and with increasing power, as the Lorentzian effects increase.

When the mass is articulated to a certain increased mass and foreshortened length (wavelength) and also suffering from time dilation, then a more slightly increase of power results in an even shorter wavelength (being done carefully); then the object becomes Noninertial, traverses space time equally at a speed of just under 3.7×10^5 of the speed of light and becomes entangled in two places at increasing distances of both place and time; thus also tunneling through not only our space time but also through Hyperspace time.

It is where a macroscopic object now is able to traverse both space and time as easily as microscopic and smaller particles do, according to quantum mechanics. The Hypersphere is the culmination of the experiments, observations and measurements of the *Time* or *Temporal Diffraction Grating*, the Hyperplane and even using the Newtonian *Conservation of Momentum Cradle* as an instrument of sorts. Also used was a Geiger-Muller *Counter*, an Atomic Clock WWVb receiver at *60 KHz* and local clocks at varying radii from the Hypersphere.

Not only are these effects of the Hypersphere local and upon the apparatus itself, but are measurable according to the *inverse square law* for electromagnetic and radiation from the Hypersphere to the inertial tidal effects following the *inverse cube law* and the slowing of local clocks at decreasing amounts at certain increasing radii from the experiments. Locales at various distances would observe how they had to reset their clocks. (Of course my atomic clock receiver would automatically update and reset with the arriving signals from the WWVb in Fort

Collins, Colorado and also the signal from the Canadian Bureau of Standards in Ottawa, Ontario, received by shortwave radio to double check.) Thus not only does the Hypersphere itself undergo very drastic Lorentzian effects, but also its environment for many miles, but with the obvious decreasing observables with increasing distance.

The Author's Personal Reflections

In the 17th and 18th centuries most of this information has been accumulated; in the latter 19th and early 20th centuries, initial stages of its more advanced geometries and even parts of its functional apparatus have existed. (In the latter 20th and early 21st centuries, it was more and more apparent to me that it was feasible. Thank God for my "odd nature" that I was never much distracted in my thinking of these third level machines, especially during these last 12 years. It was often times far too easy to contemplate the abstractions herein, while struggling to be a more convenient practical citizen of THIS time and place in human history, that I was privileged to occupy for a time.

Sir Isaac Newton said, "I have seen so far, because I have stood on the shoulders of giants". That is my situation also. I think of the times one just puts together that which is strewn about before by others. No, I do not fit into the present ambitions of human kind, so much as they drive me to solace in "God's Grand Wilderness University", something others miss because of their more fleeting 'importances' that will soon pass away with God's winds of time and space.

Again I will leave it to the reader to decide whether this is science or science fiction, and if you can disprove what you are about to read and consider.

"Times glory is to calm contending kings, unmask falsehood and bring truth to light."-- William Shakespeare.

"For a thousand years in your sight are like yesterday when it is passed and like a watch in the night." -- Psalm 90:4.

"But beloved do not forget this one thing, that with the Lord one day is as a thousand years and a thousand years as one day." -- II Peter 3:8.

Considering our fine use of celestial mechanics at the increasing scales of distance, the rocket's usefulness decreased in its limited velocities and vastly needed amounts of time to traverse space. It is fine within the inner confines of the solar system, but especially considering human spaceflight, it is far too slow. Even robotic missions at a more distant range become hindered with large amounts of time.

In 1933 AD Herbert George Wells wrote <u>The Shape of Things to Come</u>. With this advent of third level machines and with their predominantly fourth dimensional effects, the rocket is rendered obsolete, as these are "the shapes of things to come".

The Hypersphere
Journey at the Speed of Geometry
"What if there were a machine ... ?"

After concluding that all of our explorations in Quantum Mechanics, Relativistic Mechanics, along with inertial and Noninertial geometry are true (as based upon the accompanying papers), what are the implications upon any mechanism based upon these?

(Presently it is 2010 AD. This and the accompanying papers are based on 12 years of theory, research and experiment, those first seven years having been very intriguing and the last three phenomenal! If all scientific and mathematical bases are too much for many, then to comfortably consider all that is entailed as a fiction would be recommended.)

In honor of H.G. Wells' work <u>The Time Machine</u>, let us consider 1895 AD as one limit in opposition to 2125 AD, each distanced from 2010 AD by 115 years. Also the complement of spatial distance would be 115 light years along the axis of oscillation forward and 115 light years backward in time, for we are considering the mutual superpositioning of a Hypersphere in both time and space, as is proper.

Having done much historical research on my family over the years, it would have been very nice to transition from a potential to a kinetic energy state, from Noninertial geometry to an inertial geometry, from here 115 years back to 1895 AD; 'here' to mean NOW in 2010 AD as well as WHERE the earth is located as opposed to WHERE it was 115 years ago in the Milky Way Galaxy. Also what is required is an extreme resolution to 27 Market Street in Amsterdam, New York, right in front of my great-great-grandfather's bakery at a very specific time of day.

Considering the movement of not only the earth but our sun and even the galaxy's slight motion, the distance from where the Earth is today and where it was in 1895 AD is not 115 light years away, since the earth travels much more slowly than light; but it would be well within spatial superpositioning range in attempt to also superposition back 115 years.

In macrocosmic superpositioning or tunneling, which occurs suddenly in the transformation from inertial to Noninertial geometries, it entails the speed of entanglement of $3.7 \times 10^5 c$. This occurs often in the micro-, nano-, pico- and feto- scopic worlds, but rarely in the macroscopic world with which are we familiar.

In less than three hours upon the Hypersphere and with proper articulation, superpositioning can occur between 2010 AD and 1895 AD in both time and space, at the same velocity of entanglement (Noninertial geometry) and in the potential energy state (as in a tunnel diode) until a kinetic energy state in returning to an inertial geometry, while still in entanglement. Where the Hypersphere is articulated, the space and time of both 2010 AD and 1895 AD from the cellar of my home to the front of the Nicholas D Simpson's Bakery at 27 Market Street in Amsterdam, New York, are in superposition in time and space. They are tunneled together as long as the Hypersphere is oscillating.

Let us suppose that if I were to act the part of a street photographer in 1895 AD in front of great-great grandpa Nicholas' bakery (there were many street photos taken there in the latter 19th and early 20th centuries), it would not look at all suspicious. We can even consider that the cameras and films are for us today quantum information machines allowing nonintrusive, Noninertial and only potential energy tunneling. We can observe, but they only see a camera whose film will continuously carry the image information through the fourth dimension from their moment to our moment.

How intriguing of such mechanics, Noninertial and inertial geometric machines, to not only stand for some moments in front of his bakery, but to walk in the door, linger as very quietly observing customers from off the street and then quietly departing; having tried not to make any kinetic interference while entangled *here*, and *there*, in *time and space* and not breaking any known laws of physics!

In preparation I would have done a lot of prior research in the area of interest -- angle of the sun, weather as best as could be done, and even clothing and appearance of that time period. To tunnel macroscopically into 1895 AD from 2010 AD in just less than three hours and linger kinetically for maybe a half an hour, so as not to disturb the Nonlinear sequences of events, that would in 115 years have the slightest ramifications in 2010 AD.*

It is written that "we have a cloud of witnesses" in St. Paul's writings and this is true with photographs, films and other recordings from the past. But with macroscopic tunneling as hinted by James C. Maxwell's *Equations of Electromagnetic Radiation*, even as it is inertial and kinetic, has both retarded waves going forward and advanced waves going backward in time. For we again maintain the speed of light is the surface and border between space time with forward duration as we know it, and Hyperspace time with reverse time and duration. "The past" as Einstein remarked, "is as real as the future". And Gödel, in solution to Einstein's *equations*, said 'the universe should be rotating" (thus in accordance to our accompanying papers we agree) and also that we are in an expanding and rotating black hole; the other two 90° entropic dimensions of what we call 'time', the fourth dimension.

Again light and subatomic particles, even atoms and some small molecules seem to be able to pass from space to Hyperspace, forward duration to reverse duration and all back again as often as must possibly be, but without much effect at our level of existence.

So ……. what if there were a machine? And to answer that is very profound. Welcome to the *future* and to the *past*, all as real as the present NOW. It seems to only be the strength of the kinetic signal over the noise, with the strongest as NOW.

---

*The Sun-Earth traverse about one light year per every 1000 years. So for about 115 years the distance spatially is a bit more than 1/10 of a light world.

Natural Geometric Computers

Dependent on the Curvatures of Inertial Geometry

All activity, observable and measurable in the natural world, is following a greater Nonlinear function sequence and within such are smaller linear sequences. All of these functions are slow rate based on the inertia of mass for particles and momentum for photons, which are constrained by the local curvature of Inertial Geometry.

Taking a fistful of sand and depending on the local gravitational field, which is replaced here by the active Inertial Geometry of space time, the flow rate of the sand particles allowed to fall below the fist would form a pile. The rate of flow, the stability of the pile, until some sand particles would begin to roll downhill, and the width of the pile would vary under different gravitational fields. The only difference between doing this same experiment on the moon, Mars, Earth or a greater massive celestial body outside of an atmospheric consideration, is the gravitational field. The only constants are the height of the fist over the table, the number of particles and the human fist protected in a vacuum chamber to discount atmospheric effects. Thus remaining as with many pendulum experiments, the variable to our 'computer' of this single function of flowing sand is the strength of the gravitational field (the curvature of the Inertial Geometry).

Why Inertial Geometry? Because all of the particles and any photons that could be used, have mass and momentum for the particles and just momentum for the photons. Thus they all have inertia and are inertial in their nature. Therefore they are a single function and Geometric Computer. Also of interesting observation, since they are all subliminal in velocity and entropic in motion, from the more orderly fist to the less orderly pile, they are then a flow in time as well. Consider that on the microscopic scale an almost imperceptible variance of the number of grains exits the fist. A subtle variable, but admittedly important to quantum mechanics, is the Nonlinearity of the number in a rate of the sequence of events. The flow is probably not smoothly continuous as observed but has a Nonlinearity in its clusters and pulsations as well. Again quantum mechanics is near resolution. What difference is this from watching a doorway, counting the number and flow of people at a coffee shop? Not much really, though the people are more complex groups of particles and the flow is horizontal. It too is inertial based in viewing as a flow of information. The sand is a flow of information with some minor quantum variables, but essentially with just the major amount of curvature in the local Inertial Geometry of more commonly voiced gravitational field.

All of this is the flow of information. If we take Einstein's famous E=mc^2 and add "E= Information", then we are saying that energy and matter are just *information*. Seeing that according to the *Laws of Thermodynamics*, there is entropy and energy cannot be created or destroyed, and one cannot reach absolute zero Kelvin, then this is the same for *information*. It can be transduced and over time is entropic and does not ever come to a complete stoppage of flow.

$$E = \frac{mc^2}{\sqrt{1-\frac{v^2}{c^2}}} = (In)formation$$

(Incidentally, this all leads to another paper that goes far more deeply into a deterministic universe and the Nonlinear sequences of events.)

Human measurement will always be limited by its resolution, thus the often mention of randomness, chaos, turbulence, noise and chance, as well as uncertainty and probability. But within the natural geometry of the universe, any Hyperuniverse or multiple universes and even to our reference, the reverse time of Hyperspace time (time space), all is at some point beyond the lowest of human measurements and also within all the local natural Geometric Computers in our universe and any others. Even within the program of life, *Inherent Differentiation* of the unfolding topology of the RNA/ DNA computers is inertial as well. Only where the Noninertial seems to appear as the constant moving of the electron around or about a proton (or what appears to us as a superluminal like entanglement and tunneling) does this seem somewhat modified.

The natural Geometric Computers around us are but samples of the greater *natural computers* that become environmental in space and flow with time. "Eternity is ever so close and ever so far".

Inertial Geometry and Noninertial Geometry

The Replacement of Gravity and the Possibility of Faster Than Light Information Flow

A modified Minkowski *Space-time Diagram* will depict the latest results of our experiments, showing the third level machines operating as measurement and observational instruments.

Basically our universe as Kurt Gödel proposed as a solution to Einstein's *Equations*, is a rotating universe. We propose with some modifications, that our universe is a rotating black hole as other universes are. Like any other universe ours is five dimensional. The first three elements are spatial dimensions, the fourth a bidirectional time dimension, and the fifth is commonly termed 'gravity'. *Gravity* is really Inertial Geometry with a finite velocity of $c=300,000$ kms, based on the massless photon which though massless, still exhibits momentum. This fifth element, a bidirectional quantity, is Inertial Geometry and where ever our universe is in this Inertial Geometric 'ocean', its quantification is positive.

After c is super luminal (Noninertial Geometry) and is at a much higher velocity of $3.7 \times 10^5 c$, it occurs in Hyperspace time with the reversal of the bidirectional time flow, from our forward entropic (only from our perspective) to negative entropic.

This is happening much more frequently to smaller and lower momentum particles, especially light, from either side of the common event horizon meniscus that is the geometric boundary between our space time and Hyperspace time.

Access and other observation started with Nonlocality tunneling and entanglement observations in quantum phenomena and has been achieved for third level machines, not by affecting their kinetic energies, but rather their potential energy.

The tunnel diode is an electronic device utilizing electrons for this common application. The formula for the *world point line formula, $ds^2=dx^2+dy^2+dz^2-dt^2$*, with the added element dw, is Inertial Geometry at c boundary.

We used third level machines/instruments not only to observe and measure the macroscopic scale but also to manipulate macroscopic particles, especially pyrolytic graphite and bismuth, because of their diamagnetic properties for the Hypersphere, into and out of our space time and Hyperspace time, through the subtle or 'shadow' event horizons.

22

The modified Minkowski *Space-time Diagram*, slightly twisted, whole to fully indicate the fifth dimension of Inertial Geometry known as 'gravity' and the bi-directionality of time, where for us it flows positively as entropic

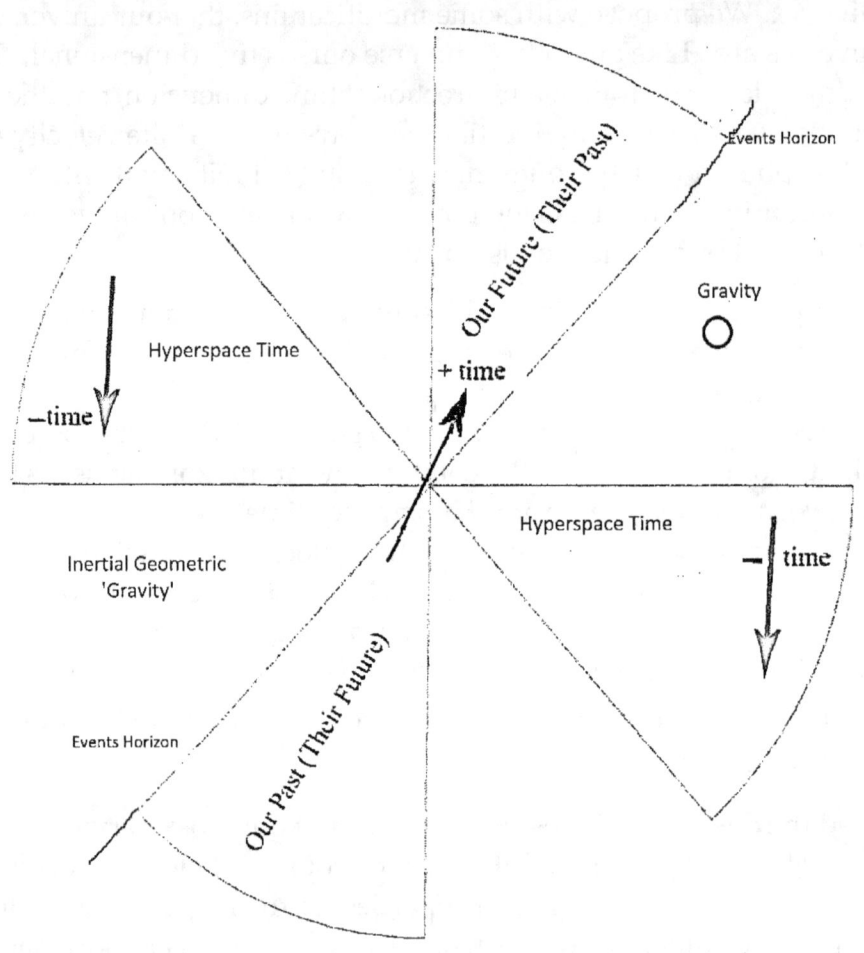

To each side of the event horizon the other side looks reversed, but is normally entropic to the inhabitants. Each appears as an 'inside' curvature to the other, each a spatial shortcut to the other. 'Gravity' (Inertial Geometry) is the fifth dimension. Time, positive and negative is the fourth dimension, and space is the first, second and third dimension. We postulate that every black hole is a portal in our universe of Hyperspace and every black hole in Hyperspace is a portal to our universe. Each black hole is rotating as well and expanding within itself.

Nonlocal Kinetic Effects across Distances in Space and Time

As per the corresponding papers in this series on Hypergeometry, Noninertial information speeds at 370,000 times the speed of light and measured effects of entanglement (Bell's *Theorem of Nonlocality* as the possible normal to our more limited observations) can now be addressed.

Given that the four dimensional surface of our five dimensional rotating and expanding universe is the speed of light itself and given that all inertial information is restricted at its fastest at *c*; then all Noninertial information is suddenly experienced at $3.7 \times 10^5 c$ *and* Nonlocality is occurring on some level at every potential energy mechanical function. Because of this Nonlocality (Bell's *theorem*) as accepted in present science, this also seems to Nonlocally connect (as entanglement) kinetic energy effects at even great distances in space as well as time.

Whether the simple pendulum or the Hypersphere of more articulation reaches the speed of light 'surface' as of a sphere (a four dimensional sphere in five dimensional inertial geometric gravity field) then the Noninertial geometry of Nonlocality is a chord. Since the early 17th century Galileo noticed the *periodicity of the pendulum* and Newton later in that century was analyzing the same. For it is in the pendulum, a machine of one moving part as a Hypersphere harmonic oscillator, an active pendulum has at midpoint its highest kinetic energy in its singular travels and its highest potential energy at each opposing end of its course; always following the *formula of periodicity of a pendulum*, concerning its mass and inertia in an inertial geometric ('gravitational') of the local inertial space time.

$$T_{sec} = 2\pi \sqrt{\frac{l,\ length\ of\ pendulum}{g,\ gravitation\ constant, 9.81\ for\ earth\ (\frac{m}{s^2})}}$$

As had initially been noticed with entanglement in Quantum Mechanics, with the Hypersphere is attainable with macrocosmic sized objects, Noninertial at 370,000 times the speed of light in both distances in space as well as time; but rather following a chord through Hyperspace time while at the point of origin and also occupying equivalences to opposing distances in both space and in time, thus protecting local cause and effect conditions. "Giving unto Caesar what is Caesar's and unto God what is God's".

Hyperspace Time, the Speed of Light and Four Dimensional Mechanics in a Five Dimensional Universe

"It is not so much WHERE Alpha Centauri is, but WHEN; and for M31 the Andromeda Galaxy even more so!" (Robert B. Cronkhite)

In our attempts of extrapolating with the third level machines for our four dimensional studies in physics, especially pertaining to macroscopic scale objects, we have been building Hypergeometric Mechanics.

We are living in a four dimensional universe that is deterministic, rotating, has event horizons and seems to be embedded in a fifth dimension of inertial geometry where momentum is an effect. After c, the speed of light is passed the Noninertial geometry takes over with no momentum. Light rides the border without mass, but has momentum.

Entanglement and tunneling observable in light, nano- and some microscopic scale particles are attainable for macroscopic particles (and we extrapolate for cosmic particles) when the potential energy is great enough to tunnel through our space time by way of Hyperspace time. When this occurs the time is reversed and so a macroscopic scale particle is able to do the same as much lower mass particles and massless photons do, with momentum being lost and Noninertial effects predominating. Near c, when potential energy is greater than kinetic energy, then entanglement and tunneling are suddenly valid.

As the coefficient of ds^2, first positive for space like, next zero for light like, then finally to negative for time like, inertial geometric effects like momentum and kinetic energy fall away to Noninertial effects like tunneling and entanglement at suddenly the speed of Noninertial geometry without momentum to $3.7 \times 10^5 c$.

Third level machines have a very predominant negative coefficient, so they are very time like and Noninertial, without momentum and with extremely high probabilistic (mechanical) distribution of potential energy which shows in tunneling and entanglement of time and space.

According to Young's *Double Slit Experiments* of the 19th century, modern quantum experiments, experiments of relativistic entanglement and now our own Hypergeometric experiments of the past several years, macroscopic objects in the right environment may tunnel and entangle. Superpositioning as well is prevalent in space and in time. If prior to our experiments a particle or photon could interfere with itself, superposition in space, now we add that it could also superposition in time. Thus an object is at more than one place at one time and more than one time at one place.

After passing the speed of light, suddenly the particle or object is at $3.7 \times 10^5 c$. It is above the drive ring containing the oscillating magnetic field, at the same time and at the same place for approximately every 1.5 minutes; it is in opposing directions one light year, also one year forward in time and one year backward in time, but Noninertial and has potential energy.

As the harmonic oscillations are articulated by the magnetic field every 2.5 hours, this distance in space and in time extends another 100 light years of opposing directions of space, and

100 years of opposing directions forward and backward in time; again all Noninertial so as not to disturb any inertial environments in kinetic energy effects, as it is only in potential energy. Its presence is locally unaware to any inhabitants, until entanglement is achieved.

When entanglement is articulated during tunneling then potential energy becomes kinetic energy and the local environment for those nonlocalities in time are affected. When the entanglement is pronounced the conditions are all remote locations of space and time, are in superposition and all can be measured and even interacted upon.

Due to the great separation of space along with time, these effects do not influence the local previous histories. This is because when *one* to 100 years of time is being articulated, the regions of space are *one* to 100 light years in physical separation. Any light signals or events that would or have been radiated in inertial geometry are irrelevant to the local historical and future Nonlinear and linear sequences of kinetic effects.

Thus Hawkins' *Causality Protection Conjecture* is so far verified. Now if the space and time separations for superpositioning, tunneling and entanglement are far more immediate, then the effects are part of the local resolution of sequences of Nonlinear and linear events, and appear irrelevant. As a matter of fact, quantumly this is happening frequently but the particles involved are so small in mass and momentum naturally, that it is considered noise. Especially the massless photon, yet having momentum, we propose it is very frequently going through such tunneling, entanglements and superpositions normally, as a timeless particle wave.

The great significance of articulating with the Hypersphere is that a macroscopic scale object transfers into potential energy from kinetic energy, first in relativistic harmonic oscillation phase where Lorentzian effects increase exponentially. Then suddenly the Noninertial, Nonmomentum phase occurs and this is just as suddenly where tunneling, entanglement and superpositioning are experienced. As the Hypersphere is functioning, these effects are expanding bidirectionally in distance and also in time, while at the same place, at the same time as where and when the experiment is operating. This is the Nonlocality that is seemingly so mysterious in Bell's *Theorem*. This is why local causality at the macroscopic scale is not observed or measured with most present non third level instruments. The resolutions of measurement are too coarse and any measurement appears as noise. As in the *Temporal Diffraction Grating Experiments* ("Where is NOW?") the inverse square law had needed to be amended by $1/\Omega\aleph^2$ for photons and $1/\Omega\aleph^3$ during the Lorentzian phase of effects prior to Noninertial entry. This is per the radii, again as noted from the point of origin of photonic or inertial information.

$1/\Omega\aleph \ r^4$ for Inertial Geometric Limited Information
$1/\Omega\aleph \ r^3$ for Photonic Limited Information

With these achievements of the Hypersphere based on earlier Hyperplane and *Temporal Diffraction Grating Experiments*, the "rocket is obsolete". The farther something is away in space, the more important is WHEN it is photonically, inertial and Noninertial of its information. And

$$E = \frac{mc^2}{\sqrt{1-\frac{v^2}{c^2}}} = (In)formation$$

The Temporal Diffraction Grating   (Abstract)

It is to be shown that time as well space can be quantified and examined by use of the *Temporal Diffraction Grating*. It is basically a "Where is NOW?" demonstrator. By this quantification of time, from his 19th-century papers, Maxwell's *Equations* imply that propagation of electromagnetic radiation goes not only forward in time but also backward, respectively retarded and advanced waves. The past and the future are as real as the present. The only difference is their kinetic signal strength to our senses and instruments of detection.

It was an experiment that utilized a laser (red of 650 nM), a photovoltaic film, clear and oriented perpendicular to the laser's beam of light; and small photo detectors placed after the laser, after the photovoltaic film and one placed before the laser opposing its output. This detector was cooled by liquid nitrogen and would detect the advanced wave in reverse time that would occur prior to initiating the laser's beam pulse temporally. This weak detection would only occur before the laser was actually fired, never before any pretense to fire.

Each photo detector was separated from the others and the photovoltaic film by 1/3 meter (about one foot). The photovoltaic film was separated from the output port of the laser by 1/3 meter as was the cooled detector prior to and opposing the laser's output port. The measurement was chosen as it is equivalent to the time light travels 1/3 meter, one nano second (*nS*).

(This experiment easily sat atop a kitchen table and is quite verifiable by any amateur to recreate.)

When observing and measuring, NOW (*present*) is assigned to the signal of the photovoltaic film, then 1*nS* forward the future detector should sense the red pulse 1*nS* later, weakened according to the *inverse law*. Then following farther down the future detector, it would sense the pulse another 1*nS* later. Already, our *present* at the photovoltaic film was the strongest signal following the laser's output. The laser's *past* was our *future* and our *past* was the following detector's *future*; then readings in the *future* were based on the reception of the red laser pulse from the just then detecting sensor. So the *past* was prior to the laser and the *present* was after the laser.

Now the *past*, prior to the laser was the detector in opposition and before the laser. It was cooled by liquid nitrogen so as with the least thermal noise interference could detect an advanced wave in reverse time as Maxwell's Equations implied. The strength of this wave is very weak and seems to be following the linear attenuation linearly as per the *inverse square law* but severely truncated. It is more applicable to apply the Omega Aleph as a factor to multiply the denominator, as in this form: $1/r^2(\Omega\aleph)$.

Just before the initiation of the laser, without any pretense there is a signal as per detection in reverse time at 1*nS* prior to the laser's pulse initiation. The signal is detected as the laser's output is preceded by the pulse radiating before the laser fires by 1*nS*.

With our tabletop *Temporal Diffraction Grating* we are able to quantify the fourth dimension in not only forward time but in reverse time as well. We are able to measure, observe and quantify that the *past* and *future* are as real as the *present* (NOW). It is just a matter of kinetic signal strength over noise. The strongest kinetic signal is the *present*, then the *future* following the inverse square law; before the *present* and far weaker is the *past*. Quantifiably, the *present* is after the *past* and before the *future*; light is on the border between forward entropic time flow and the apparent reverse entropic time flow of negative time. It is only relative in what appears to be the negative time flow to us, is normal forward entropic time flow on the other side of the border of the speed of light. For on the other side of this light border is Hyperspace time; light is actually on both sides of this border, for light marks an event horizon. (Another paper explores the function of the Hyperplane.)

Nonlinear Sequences of Events   (Abstract)

The purpose of this paper is to propose that the sequences of all events in the universe especially locally, are deterministic Nonlinear sequences that can be quantitatively and experimentally verified, from the quantum scale, to the macrocosmic scale and to the cosmic scale; that human free will is limited to choice selections --- what appears to be random, chaotic and turbulent occurrences --- that are quantifiable and actually microstates in greater macrostates, as well as reoccurring in a far greater expanse of time.

The use of third level machines that have predominance in time like regions was required here. The *Temporal Diffraction Grating* and Hyperplane were utilized. Later the Hypersphere was oscillated and the effects noted in what appeared to be games of chance at the Turning Stone, Borgata and Tropicana Casinos in Atlantic City, several casinos in Las Vegas and the Casino de Montréal in Québec, Canada.

Extrapolations of these Nonlinear sequences were also utilized in the Kentucky Derby of 2008 AD. All games, activities of humans visiting a coffee shop, weather patterns and solar events such as sunspots (quiet and active solar occurrences) and even historical and political activities were noted. In all occurrences, what appeared to be random, chaotic and turbulent and even benign activities (customers purchasing coffee at a local shop) were fitting into longer term patterns of Nonlinear sequences of events.

By telescoping perspective to minimize human activity and even games to entropic machines and seeing them as if quantum like in scale, then all events and human activities were easily likened to a microscopic perspective. It is written that "we choose between two masters" is essentially the summation of these quantifications

(In a prior paper on third level machines and inertial geometry it is seen that all activity is deterministic with the *past* affecting the *future*.) The Nonlinear sequence of events proceeds along the fourth dimension and is limited in allowance of decisions with real outcomes with only those of highest probability as in quantum wave function collapse. The probabilities are at first quantifiable but later with inclusion of Poincare's *Reoccurrence Theorem*, the quantities seem to congregate at $x^e \times 10^{40}$ of high probability of reoccurrence. This is longer than the known age of the universe.

The greater implication in respect to the third order machines is that fourth dimensional effects are very apparent. The fourth dimension along with inertial geometry is a basic matrix upon which all else is fluctuating.

In the use of the *Temporal Diffraction Grating*, where it is quantified in agreement with Maxwell's equation's implications of advanced and retarded electromagnetic propagations, that *past, present* and *future* are each as real as the others, while the strongest kinetic signal to noise ratio is with our perceived 'NOW'. Less weak is the *future* following the *inverse square law*; the *past* as prior to NOW is very weak. *Past* prior to *present* is the *inverse square law* divided by Omega Aleph, the multistate prime number constant of inertial geometry (in *'Omega Aleph the Multistate Constant and The Prime Number of Primes'*).

So what is derived from all these quantifications is that deterministic Nonlinear sequences of events in our *local* universe and *globally* in our universe's greater scale is the following of *present*, then *future*, and *past* very weakly, prior to NOW formulations.

$1/r^2 (x^e \mathbf{\times} 10^{40}) \mathbf{\times} \Omega_\aleph$ for the future of the 'NOW signal'

$1/r^2 (x^e \mathbf{\times} 10^{40}) \div \Omega_\aleph$ for the past and prior to 'NOW signal'

Human choice and the possible sequences of events as per the fourth dimension are deterministic with greater probabilities indicating what is in reality beyond most resolution of measurement certainties.

The scale from personal to nation states and the events in human history are also following deterministically the entropic information flow, with very limited choice allowances at all points. Again we have the responsibilities of what we decide between our "serving either of two masters". There is always a finite number of states in the infinite. They are complex microstates and phase spaces within a greater volume of macrostates and phase spaces, all well incorporating the fourth dimension and inertial geometry of space time.

A Practical Exhibition of the Fourth Dimension in a Five Dimensional Universe and
The Formulation of the Rate of Nonlinear Sequences of Events in the
Fourth Dimension, Inertially Based

Here is the practicality of the presence of importance on a scale that is easily perceptive on an *anyday* for *anyone, anywhere*.

First I approached a small group of people by saying, "How is everyone *today* and right *at this moment*?" Playing along of course all offered their respective valuations. After this I exited back through the doorway from which I had just entered the room. Waiting a few minutes I reentered the room saying, "How is everyone *now*?" and again everyone offered their current status. I then said, "Wait a few more minutes and I will see you in the *future*!" again leaving the room. After a few more moments of this theater, I reentered saying, "*Now* I am meeting you in the *future*!" which of course was the *present* as well as the *future* of the previous NOW in our second conversation.

Though this seems a bit theatrical, it was a way to open the observation for them of what is most often not consciously observed. It was the quantification of the sequence of events of a certain place along a quantitative piece of the fourth dimension. I did see them again in that *future* after that *present* NOW, which was also in succession of that first *past* moment when I opened this 'play'.

The quantifications of my 'enterings' and 'exitings' were not very accurate based on the four times elapsed in this trial of events. This made the sequence more Nonlinear in time, while my arrivals and departures (just me) were linear in being just one element. If I would have spaced all arrivals and departures accurately to an atomic clock and the time equally spaced for every four minutes, it would have given this theater a very linear sequence to events portrayed.

In a more realistic complication we reenacted the play. First I came in and left after four minutes, then returning and leaving every four minutes. But with each successive arrival and departure after the first, one person exited and returned with me; then successively two and then three, increasing the complication of the number of elements arriving and departing every four minutes. Now we had a quantified linear time base with Nonlinear elements arriving and departing. We maintained a time record like a log and entered the number of elements at the time they arrived and departed.

This was now an accurate record of the real world's Nonlinear sequences of events, the normal experiences of life along the fourth dimension. The only consistent element was the accurate marking of the time and the elements coming and going with Nonlinearity.

It is this Nonlinearity that is on the grander scale of time able to exhibit greater patterns of cyclical occurrences. If the comings and goings of a day at a lunchroom were logged, one would see the apparent chaotic Nonlinear elements of events in the comings and goings to greater patterns of clustering of the days' breakfast, lunch and dinner times. Also for a year one would notice patterns and occurrences associated with the yearly cycle of holidays.

What at first appears chaotic, random and turbulent is really on a grander scale of the fourth dimension reoccurring, though not exactly Nonlinear sequences of events. Even on a much grander scale than in Poincare's *Reoccurrence Theorem* to a greater time period of the universe, $n \times 10^{38}$ seconds It is still entropic even with reoccurrence but this arrow of time on our side of Hyperspace time would still progress according to the *Second Law of Thermodynamics*. The universe as we conjecture is also deterministic with choices allowed in less areas of time than the longer deterministic areas of time.

Next with the Hypersphere in oscillation locally, we use two simple instruments of Sir Isaac Newton's invention. Two Newton's *Cradles* are aligned axially and separated by several meters. The longer the separation of the *cradles*, the proportionally more sensitive the next measurement of the local space time curvature. Pieces of paper are placed between the spheres of each separated *cradle*. The Hypersphere is kept in Lorentzian effects mode by keeping its activity inertial. Another dual set of Newton's *Cradles* are set up several kilometers away and another pair even farther away. Each *cradle* is aligned perpendicular to the Earth's magnetic field so that the sphere does not become magnetized as it moves according to the changes (in soon to occur) bending of local space time. This will also affect clocks locally since clocks are passive and will be periodic depending on this 'bending' of local space time. The bending is an oscillation and over the time of this experiment the clocks will vary from normal to slower rates of operation; this will be occurring in all activities of the sequences of the events locally too. No one will notice unless they later compare their clocks to an atomic clock receiver. This is a commonly purchased device and receives 60 KHz signal from the WWVb at Fort Collins in Boulder, Colorado. The local atomic clocks will catch up to WWVb after their environments locally remain at the returned normal space time curvature.

The Hypersphere is also an active clock because its inertial Lorentzian oscillations bend its local space time. And according to the *inverse cube law* the local space time is around $1/r^3$.

These are tidal effects and the slowing of the clocks' time dilations are able to affect the rates of duration of the surrounding environment of the Hypersphere. Thus the local resulting fluctuations of the local space time geometry ('gravitational field') is what we conjecture in other papers as Inertial Geometry.

When the Hypersphere slips into Noninertial geometry mode, it suddenly is at $3.7 \times 10^5 c$ and in superposition of space and time and entanglement; then these inertial Lorentzian effects also subside. This is when the Hypersphere is at a potential energy, not kinetic and is also tunneling while Noninertial and no longer able to interfere with its environment, locally or Nonlocally.

Back to the effects of the Hypersphere in its increasingly Lorentzian disturbances on the periodic dilating of clocks and all other cyclical activities, including the rate of the sequences of events and duration locally, then these effects are also noted in the behavior of the respectively spaced Newton's *Cradles* following decreasing intensities according to the *inverse cube law* $1/r^3$.

The pieces of paper between the spheres begin to fall earlier and more suddenly the closer the *cradles* are to the Hypersphere. On a table one could set up a Newton's *Cradle* and with each person on their tabletop noted on the map, each could record when their *cradle* lost its paper inserts.

An idealized map with distances interposed at every tripling of the distance from the Hypersphere's oscillation

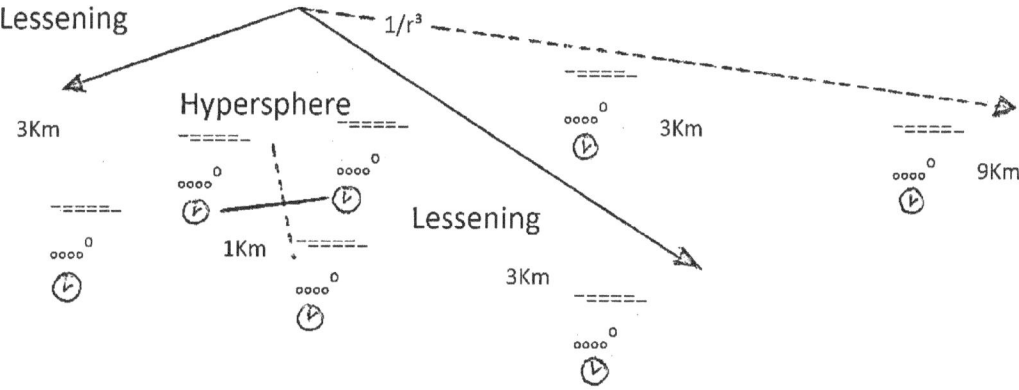

In our rotating, expanding Gödel black hole universe which is deterministic, having above the speed of light surface, forward entropic time flow and below the speed of light surface, reverse entropic time flow, we are in space time, above and inertial. Hyperspace time is below and also inertial. So we both share the one side of the two directions of the fourth dimension. Our initial inertial geometries are the positive side of the fifth dimension or Inertial Geometry.

When either side has a particle (mostly microcosmic scale) tunneling, that particle has transformed most of its kinetic energy into potential energy, can be entangled and experience superposition of space and superposition of time. It no longer has much kinetic effect upon its local environment and is very Nonlocal in space and in time. With the instruments or machines of the third level where the coefficient of ds^2 is negative and time like predominantly, then it seems no longer here or no longer, just in here in space and time, on either side as noted in our experiments and observations. This conjecture with our third level instruments and machine profoundly extends to the common scale of our everyday experience, macros cosmic objects and the Hypersphere.

The use of Newton's *Cradle*, the *Temporal Diffraction Grating*, the Hyperplane and the Hypersphere allows not only observation and measurement, but also articulation from kinetic to potential energy states, even for an object of everyday scale to tunnel as electrons do in tunnel diodes and experience superposition in space, time and entanglement.

As the universe is finite, this would be the formula for the rate of nonlinear sequences of events in the fourth dimension.

$$2\pi \sqrt{\frac{\Omega_{\aleph}[matrix_{n=\infty}^{n=100}][[P(E)=\sum_{j=1}^{n}P(E_j)P(E/E_j)](8)(10^{40})]^e}{G}} \bigg/ \sqrt{1-\frac{v^2}{c}}$$

Many "cannot see the forest for the trees" and thus are not at first able to observe the fourth dimension; but with these demonstrations for everyday experiencing they see it. It was King Solomon who noted, "There is nothing new under the sun", for *this all has been here all the time.*

Approach, Translation and Passage through an Event Horizon

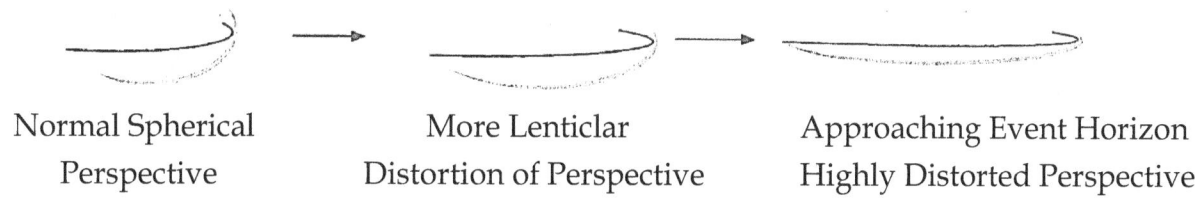

Normal Spherical Perspective   More Lenticlar Distortion of Perspective   Approaching Event Horizon Highly Distorted Perspective

Above is a simplistic representation of distorted Lorentzian effects of the perspective from a distance near a Hypersphere's articulation in approaching an event horizon. From a distance it would appear to be spherical, then more and more lenticular, to flat disk like perspective as it approaches the Event Horizon. The perspective from within, the outside to greater distances would also be a surrounding spherical to lenticular to flat disk like observation. Lorentzian effects of mass increase, foreshortening the length in the direction of oscillation and time dilation would increasingly be predominant.

Nearer the Hypersphere, above the sky would darken; yet because of the nearness of the floor/ground adding to the distortion along with the Hypersphere, below would look normal. Unless one was observing from a great height (a tower) then below the ground would darken. This would be the effects of the swiftly changing red and blue shifts from the oscillations occurring at very high frequencies in the Terahertz range.

With the Hypersphere viewed as if through a window, the outside world would appear as a graying but bright ring with darkness above and below. Quickly the grayness would blacken then return and the view would again go more lenticular and then spherical as oscillations decreased. From the outside at a distance one would see the Hypersphere expand from a reddish/bluish/grayish disk to being more lenticular than spherical.

If the Hypersphere is allowed to articulate locally through or translate an Event Horizon, its passage would be more intriguing.

For the outside observer would see a fading away of the Hypersphere as it slips through the Event Horizon, disk like and ever darker. Now entangled in two places at once, as well as two times at the same place, both times and places equidistance from the time/place of origin, in opposite directions; the Hypersphere in our space time, past and faster than on this side inertially and in Hyperspace time Noninertially.

Within the Hypersphere, looking through the window, when articulated at entanglement one would see what appears to be our universe (which it is), but on the other side of the Event Horizon and in great distance of space and time; but to us it would appear in reverse entropy, negative time, so as in reverse duration. (The distance equivalence of space and time are covered in a previous paper.)

During the 'instantaneous translation', Noninertial Geometries predominate. Also from our perspective Hyperspace time inertial effects are to us Noninertial, like potential energy. Inertial Geometry effects are only inertial on either one side of the Event Horizon or the other, where they are kinetic energy. Thus, reverse duration events have no effect on forward duration events, they are only potential to the other if the decoherence occurs as entanglement ends, for an object is now kinetic on only one side or the other of said Event Horizon. So when an object is on either side its kinetic relationship is limited to that side. When in entanglement during translation and passage, it is then potential on its side of origin and now being on the other side is kinetic only on that side along with that side's reverse duration and Hyperspatial location. It is *Hyper-spatial-temporal* and kinetic there now.

After articulating an Event Horizon and the object is local in our space time, then its reappearance above the oscillator disk fades back from disk like to lenticular to spherical again. It is decohered to our present space time --- returning from Nonlocal, potential to kinetic energy, from reverse duration and entropy to forward duration and entropy, from some distance great or small, from bidirectionally in time past and future, and bidirectionality of space above and below the oscillating disk. As such, during translation because it is potential and not kinetic, there is no interference or 'collision' with any objects on this side of the Event Horizon. On the other side when translation from potential to kinetic, then interference/collision is possible.

It is written, "Wisdom is justified by her children" and so is Hypergeometric Mechanics.

Omega Aleph, $\Omega_\aleph$ the Multistate Constant and the Prime, Prime Number

It is commonly presumed that *one* is not a prime number. The 'rule' that a prime number is only divisible by itself and *one*, is politically denied to *one* and so for most of Western mathematical history *one* has NOT been allowed to be a prime number. Even children in school ask why *one* is not to be considered a prime number when it meets all of the requirements.

But we welcome *one* as a prime number and soon to be seen as the *prime*, prime number. It is the border of the linear and Nonlinear. It is just *one* and it is also more. As a matter of fact extrapolating about *one* and discovering implications of *one*, leads to the profound revelations that it seems quite key to the concepts of quantum mechanics, relativistic mechanics, space and time, *zero* and *infinity*. It becomes more apparent that they are all shades of the same thing.

We propose that *one* is the *prime*, prime number, is a multistate constant and also a super positioning constant. It is also equal to *zero*, *infinity*, the squares of *zero* and *infinity*, and the square roots of *zero* and *infinity*. It is equal to the square of +1 and the square of -1. It is also amazingly sometimes equal to *zero* and *infinity*.

All shades of squares and square roots = $\Omega_\aleph$

$$1 = \frac{0}{0} = 0 = 1 \qquad \frac{0}{1} = 0 \qquad \frac{0}{\infty} = 0 = 1 \qquad \frac{0}{-1} = \Omega_\aleph \qquad \frac{0}{\sqrt{-1}=i} = 0 = \frac{0}{i}$$

$$\frac{1}{1} = 1 = \frac{0}{1} = 0 = \frac{\infty}{\infty} = \infty = 1 = 0 \qquad \sqrt{-1} = i = \frac{\sqrt{-1}}{\sqrt{-1}} = 1 = i = 0 = \infty = \Omega_\aleph$$

$$\frac{1}{0} = \Omega_\aleph \qquad \frac{1}{1} = 1 \qquad \frac{1}{\infty} = \Omega_\aleph \qquad \frac{1}{-1} = -1 \qquad \frac{0}{\sqrt{-1}=i} = \Omega_\aleph = \frac{1}{i}$$

$$\frac{\infty}{0} = \Omega_\aleph \qquad \frac{\infty}{1} = \infty \qquad \frac{\infty}{\infty} = \infty = 1 \qquad \frac{\infty}{-1} = \Omega_\aleph \qquad \frac{\infty}{\sqrt{-1}=i} = \infty = \frac{\infty}{i}$$

$$\Omega_\aleph = i = \sqrt{-1} = -1 = 0 = 1 = \infty = \Omega_\aleph$$

This can even be illustrated with a Venn diagram. If we rename *one* as Omega Aleph, $\Omega_\aleph$ then we see *one* easily as a multistate and supporting constant; along with *one*, -1, *zero* and *infinity* are shades of the multidimensional in space time and Hyperspace time. As the square root of *-1* it is thus equal to the imaginary number which is $i = \sqrt{-1}$! It is the number in Euler's exponent of *e* in his famous Euler's *Identity Formula*:

$$e^{i\pi} - 1 = 0$$

The Venn Diagram of Set Theory

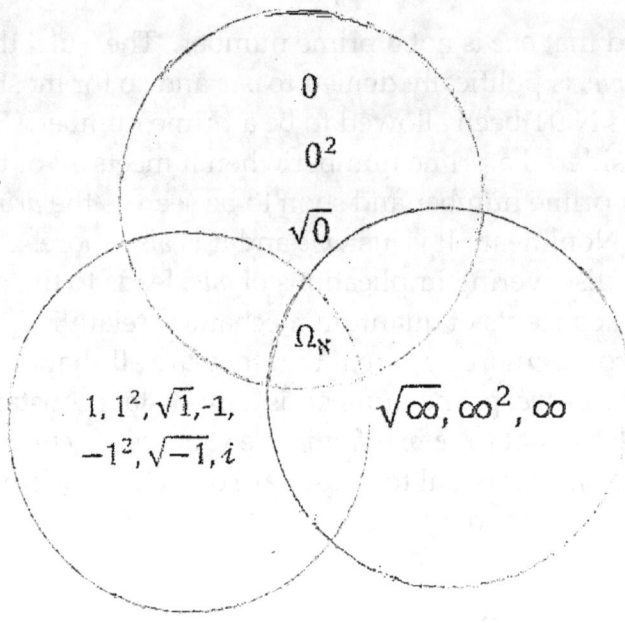

In Hyper geometry, inertial and Noninertial geometry it is better to understand the third level machines that are temporarily predominate in function and to understand four dimensional mechanics, Hyperspace time and our space time.

It is very significant that in the Hebrew *one* as singular is *echat*, while also in its compound concept *echad*!

$$e^{i\pi} - 1 = 0$$

Omega Aleph, $\Omega\aleph$ Locations upon a Riemann Sphere

Effectively locating Omega Aleph, $\Omega\aleph$ upon the Riemann *Sphere* is locating "Where is WHEN?". We are applying this predominantly four dimensional view of space time (time space).

Following are two very similar figures illustrating Bernhardt Riemann's *concept* from the 19th century of his famous *sphere*. After the placement of our proposed multistate, super positioning constant, $\Omega\aleph$, a quantity of superposition upon not only the *Real Number Line*, but also the *Imaginary Number Line* including *zero* and *infinity*. As is with chaos, turbulence and random chance, the concepts of *zero* and *infinity* are our way of saying that something is beyond our resolution of measure.

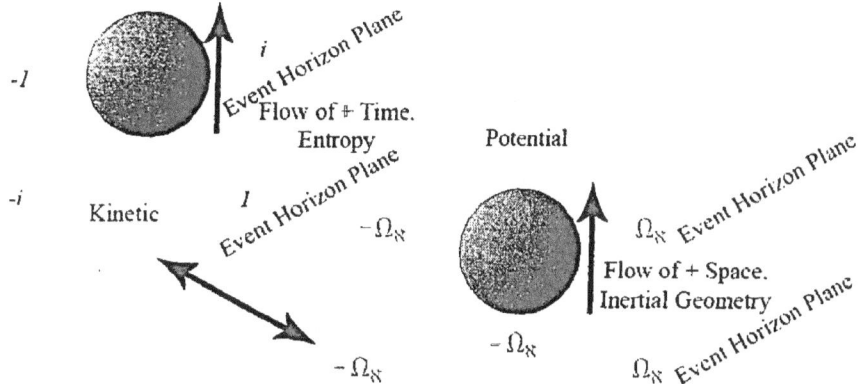

In a predominantly four dimensional view WHEN is a *place*. (This is a parallel concept to my other paper on the "Where is NOW?" experiment with the temporal diffraction grating.) On a Riemann *Sphere* space is replaced by time; space is of a single dimension with time as not only one but actually three dimensions!

But at the present level of progress we will focus on the major entropic time that we are familiar with, yet keeping notice on the other two entropies of time upon the four dimensional Riemann *Sphere*, as is being proposed in these papers. Each side of any Event Horizon has this same nature, with one major difference -- the flows of entropies are in opposition from one side to the other.

This is of the utmost importance to not only time symmetry: T: $+t \rightarrow -t$ but to the symmetry of potential and kinetic energy: E: $+PE \rightarrow -KE$ (Potential energy as per a pendulum the inertial opposite of kinetic energy.)

$$E = \frac{mc^2}{\sqrt{1-\frac{v^2}{c^2}}} = Information = \frac{Potential\ Energy_{Non-inertial} \leftrightarrow Kinetic\ Energy_{Inertial}}{\Omega\aleph}$$

Essentially in a predominantly four dimensional environment a potential energy state can be in superposition anywhere in time or space. In a kinetic energy state there is sudden incoherence to a present inertial time and space. Implications from the Riemann *Sphere* -- from abstraction of real versus imaginary numbers to practical applications for inertial to Noninertial, kinetic to potential energy, incoherent to single time and place, to the super positioning to many times or places -- are of practicality for not only quantum, microscopic and smaller objects, but also for larger macroscopic and cosmic objects. Thus the Hypersphere and its predecessors, the Hyperplane and the *Temporal Diffraction Grating*, are now far more conceptual to understand and practically articulate.

Again all inertial kinetic activities are limited by the speed of light, for even light has inertial momentum while having no mass. Light is the four dimensional surface marker. It is easier for light than microscopic particles, because of very little momentum; then with the proper articulation for the local environment, macroscopic particles are able to transfer to Noninertial potential activities at the speed of near 3.7×10^5 times the speed of light suddenly. Thus, WHEN is far more important than WHERE.

It has been written of things to come as if they already are and the past as a present moment. King Solomon said, "There is nothing new under the sun" (as well as our galaxy and our universe). Even Poincare's *Recurrent Theorem* alludes to these things.

Addendum

By rotating the Riemann's *Sphere* representation by 90°, a second dimension of time appears, proposed here as rotation; with a third and final rotation by another 90° transformation we arrived at another dimension of time, proposed as an expansion. This would be illustrated provocatively as Gödel's *solutions* to Einstein's *equations*.

Here we propose three dimensions of time -- the familiar entropic arrow of time, the rotation of the universe (increasing very slowly, also entropic) and the increasing expansion of the universe, also entropic. Space is reduced to only one dimension (following on the next page is an illustration of this).

We propose that the universe is a rotating, expanding black hole with opposite symmetry to space time with Hyperspace time. Also that the speed of light at the 'surface' or border between these two regions, inertial geometry, an active kinetic geometry, with Noninertial geometry as the potential geometry. Time is reversed on each side within the region; space or Hyperspace are appearing as strange and opposite to each other, only relative to the other. But each is the same from within --- all normal. Each sees itself as the outside convex surface, rotating and expanding with opposite symmetry and their opposing entropic time.

3 dimensions of time at 90 degree transformations with space as only 1 dimension

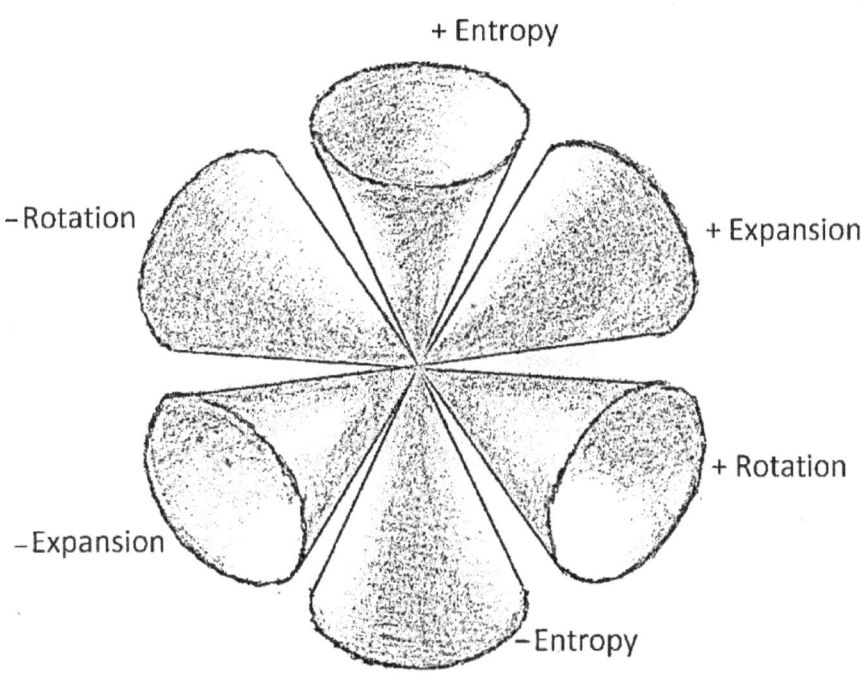

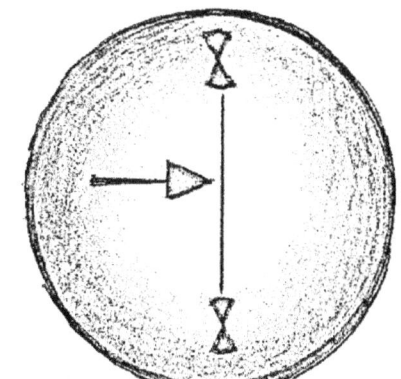

Here time has 3 dimensions and space only one!

Asymmetric Evidences in Measurements of a Rotating Universe

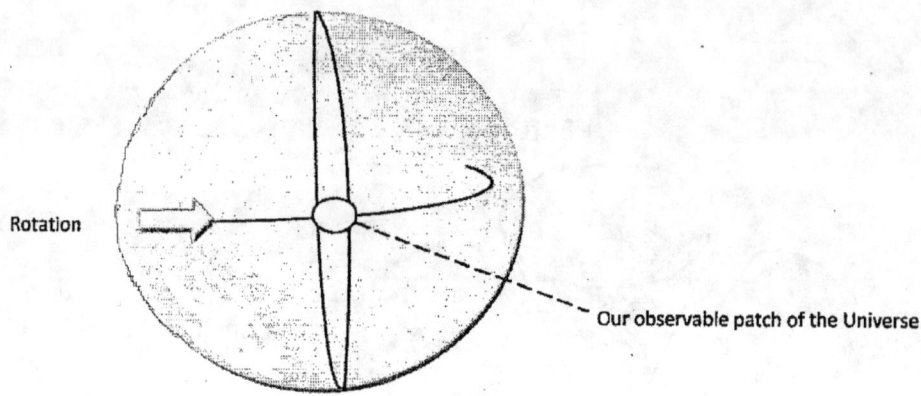

An asymmetric discrepancy should appear in very great resolutions of measurements of element to element abundances, the *Finite Structure Constant* and the appearance of the surrounding galaxies' movements through space; particularly from our measurements in angles of oscillation of the Hpersphere and angles of oscillation in reference to a laser light upon the Hyperplane.

The Hypersphere is limited in its use for any extreme angles because it is dependent upon a very direct perpendicular position to the local 'gravitational field' (local curvature of Inertial Geometry) in order to oscillate correctly.

This asymmetry should be observable at any patch of four dimensional space time, which in the above illustration is compressed to a two dimensional surface equally at the speed of light at the Universe's Event Horizon surface. Below c is just above the Event Horizon's surface as per speed relative to So on this side of the Event Horizon we experience forward duration and positive entropy, below light speed Inertial Geometric effects; we see the universe expanding, note the differential abundances of elements in one direction to another at great scales of space, as well as the appearances of constants --- like the *Finite Structural Constant* (a), as though space itself is showing a 'Doppler –like' shift.

In 1947-48 Kurt Gödel asserted a *Rotating Universe* in his solutions to Einstein's *equations*. (As brought out in another of my related papers, the universe is a rotating, expanding black hole, four dimensional in space time and in a fifth dimension of Inertial Geometry; and we are on the surface.)

To reiterate, our surface is attained at the velocity of the speed of light, approximately 300,000 km/sec or around 186,282 mi/sec. Below that speed Inertial Geometry predominates until the speed of light. At c there are turbulent Lorentzian effects where a quiet Noninertial Geometry takes over in Hyperspace time all appears 'normal' where entropy is increasing, yet for us it is reversed, space is expanding and the universe is rotating.

From either side of the Universe's Event Horizon above c, with reverse duration/reverse entropy in relation to our space time ('normal') point of view. In Hyperspace time or any Event Horizon, the region to the other side looks as a sphere from its outside going in, a positive curvature. For we propose that the Universe's Event Horizon and all smaller ones are actually the same place, thus appearing as Nonlocal. So by carefully measuring constants and abundant differences of elements, and observing the 'flow' of the galaxy's motion, especially at greater distances, the rotation of our universe is detectable.

It is written, "The heaven for height and the earth for depth and the hearts of kings are unsearchable." So we can surely not understand with our ever constant limitations of resolutions … yet in God's great tabernacle of the suns and worlds within, He does allow us glimpses.

Inherent Differentiation on a Galactic Scale

Galaxies as Biometric Systems for the Topological Unfolding of RNA/DNA Natural Computers

Here we are looking at life as we know it, a biological sample of a greater galactic system. Because we do find this sample then it may be extrapolated that it is at some level prolific enough to have at least this one known complex and long term ecosystem. Considering life as a mathematical statistic helps to quantify it, thus the geometric theory brought forth here is that life as we know it, is a geometric function.

Inherent differentiation is the progressive topological unfolding of a programmed Nonlinear sequence of the hard/software RNA/DNA computer, triggered only by environment and time. The initiated program at this point of human based science is still lost in space and time. Within the space time of our possible four dimensional universe (that may be a rotating black hole which has been slowly decreasing in its initial fast rate of expansion), when it possibly was born within a large stellar collapse of a greater universe or Hyperspace. It could be that Hyperspace time (time space) and our known universe are within an even larger five or more dimensional space time. There may also be a second time perpendicular geometric to the forward time in our universe and the reverse in relation to us time of Hyperspace. (This is covered in far more depth in another paper dealing with Inertial Geometry.)

This Inherent Geometric progressive program herein described is further postulated to be *common* within our Milky Way Galaxy and at least similar spiral galaxies including Andromeda (M31), Triangulum (M 33) and possibly some of the other less similar galaxies that may have some galactic locales of the proper environment for this progression to unfold.

Given such an Inherent Differentiation computation that is continually active even when appearing dormant, then life as we know it and any other life with Inherent Differentiation matrix may indicate that life is prolific. As seen in the geometries of nature, the complex and larger are rarer in experience than the more abundant, smaller and less complex. This is seen in the animate as well as the inanimate. Considering the extreme hardiness of the less complex, that would indicate by extrapolation that they are more prolific and would be the base upon which the more complex and rare, next level complexity would unfold.

Thus with the universe as a grand computer, life would also need to acclimate in parallel in order to survive and spread. As William Shakespeare said, "Though this be madness, there is method in it"; and in agreement with Albert Einstein, God does not play dice, for He does not need them (*only humans do*, Robert B. Cronkhite).

Earth as a Prehistoric Water World

This is an exploration of the possibility that at one time in the earliest history of our planet, Earth was a water world, a super planet of possibly two, three or more present Earth masses. Its atmosphere at that time would also include hydrogen and helium in its upper atmosphere and much more water. There would be no continents and about three times more the gravitational pull.

This would easily have allowed more thermodynamics in Earth's weather, ocean currents and even organic chemical reactions. The much thicker atmosphere would be very protective of the global ocean's surface from any cosmic radiations, even from any nearby supernova or possibly closer gamma ray bursts as well. Complex life's inherent differentiation in the possible DNA/RNA topology programming would be better able to take its progression from the continual triggers within the environment and time to safely unfold according to quantum superposition with that biogeometry. This macroscopic wave function unfolding would have been preceded by the same on the microscopic scale in time.

If we take the lower estimate of two Earth masses -- then with a little more gravitational pull, small outer hydrogen and helium atmospheric envelopes and more sunlight allowed to the surface than a three dimensional Earth mass model, life would still be quite possible. A three Earth mass model would have more atmospheric pressure, less sunlight arriving to the surface and more hydrogen and helium in the upper atmosphere. Again staying with our 'two Earth mass' prehistoric Earth, photosynthesis is more possible for chlorophyll plants as well.

Considering Earth's present orbit around the sun it could have formed either closer or further away; yet with our two Earth mass model, its temperature stability would be quite constant -- with a little thicker atmosphere, water vapor, CO_2 and probably CH_4 for greenhouse gases acting as temperature stabilizers for consistency over long periods of geological time.

Adding to later interplanetary impact to form our moon would have been catastrophic, but less fatal than for Earth at its present size or even just slightly larger. Many exoplanets that are recent discoveries, especially water bearing ones, seem to be rather numerous and so far are around the more easily discernible environs of M – dwarfs stars.

Also with so much water and a quite deeper ocean globally than we presently see, our planet with our hotter G type star sun (even considering around one third less solar output), again Earth based oceanic complex mega flora and mega fauna would thrive.

Returning to an early interplanetary impact for our moon's creation would have caused a mass extinction easily, but not nearly as totally as for our Earth at its present, slightly larger size. Thus with its great water shield around it as a protective cloak from space radiations and particles, but also a far better recirculation of uplifted material and debris from the impact to keep from leaving the Earth. Assuming there was much material lost into space, some would have remained in orbit and would eventually coalesce to form our moon, which has been considered originally to be made up of much uplifted and strewn mantle material. With the loss of mass and the great kinetics involved, the loss of the hydrogen and helium upper atmosphere, and with less water, oceans would still be available; tectonic activity, that under greater pressure of atmosphere and much deeper oceans would have been possibly subdued more, much of this tectonic energy would have been released. Yes this would have still been a very devastating mass extinction for most of the mega flora and mega fauna, and quite a bit for the microscopic life forms also.

Considering the RNA/DNA as a software and hardware computer that has within it the possibilities of programmed, unfolding, topological biogeometry, then with the continuing more consistent environment and time triggers, this unfolding would again occur in its progression. Speculating with the earth as one example of many (and since exoplanets are being discovered more and more lately, including super Earth water worlds), then this may on cosmic scales of many spiral (barred spiral) galaxies like our Milky Way Galaxy, is statistically over deep time quite normal even if on an extrapolated estimate of 10% of sun like stars with planets such as M, K, G and F types. Further speculating that RNA/DNA as complex as it is, carrying such inherent differentiating biogeometric programming, needing only to be triggered by a consistent enough period of environment and long enough time, then life as we know it is quite prolific on the scales.

As St. Augustine remarked, "God is the maker of worlds" and He has sure made many of them, as we are discovering more and more each day.

Helium Burning in the Sun and the Case for a Solar Nova within Recordable Human History

As for most of recorded human history the sun's activity has not been noted, and by inference only from tree rings, ice coring and other indirect manners, understood. It is within interest that possibly low level helium flashes, products and other indirect observations have reflected the beginnings of helium burning within the sun. In most observation the solar radius has been shrinking. There seems to have been some perturbations of slight expansions upon this radius and also slight apparent byproducts of helium flashing. Helium burning is not possible with a mass of our solar type star, but helium flashing is well within the parameters of the present science of stellar evolution for our sun.

The possibility of a solar nova, even a slight one, would be of extreme importance to human civilization and life on Earth as we know it. The effects upon the Earth's atmosphere, surface radiations, winds and tides would be of enormous consequences. By many scientists' calculations these effects are not to be expected for another one billion years, give or take 500 million years. Since it is speculated to be that much, it has to be extrapolated before human measurement and then we may also be off by another 500 million years or so.

The effects of a helium flash or a series of them would be of great significance to our present human endeavors and life at present. The greatest of changes would not be immediate, but their cumulative effects would continue; more complex life would be the most sensitive and the less complex would follow. Weather, winds and even gravitational tidal effects of the sun would vary observably. Sunspots would not be very prominent at times, then becoming very prominent, but not following the 11 year cycle. And it is quite assured that the governments of the world would hinder the truer revelations as to belay panic.

It would be a grand humbling of human endeavors when the sun becomes inconsistent and even more dangerous than has been known in recorded human history. The resulting mass' losses from the solar surface would affect the upper atmosphere and yet from nearby stars it would look quite lovely as the multicolored gases were given off. Our demise would be their scientific study.

It has occurred before … the great delay in sunspots and the resultant solar activities emanations of particles and electromagnetic radiations. But considering it has occurred before and even by inference from tree rings, ice cores etc. also in recorded history by our great scientific observers of the past, and for us of the just past solar cycle, could we be 'seeing' the beginning of helium flashings?

The results would be catastrophic if a large helium flash occurred. It would not only change life on Earth, but humble mankind to oblivion.

"Knowing we had the knowledge and gifts, and wasted them would be among the greatest regrets." (Robert B. Cronkhite)

www.ingramcontent.com/pod-product-compliance
Lightning Source LLC
Chambersburg PA
CBHW081620170526
45166CB00009B/3041